The Number System

By: Ramesh Kumar Roy

B.Tech, IIT, BHU

DEDICATION

Dedicated to my father who made me aware of the wonderful world of
numbers and who is shaping a lot many talents even today as a dedicated
teacher

CONTENTS

ACKNOWLEDGMENTS

I would like to express my gratitude to the almighty lord first and then to many people who made me write this book; to all those who provided support, resources, their valuable advice and encouragements.

I would like to thank my father as all the education and writing has originated from him and also to my beloved school Netarhat School, Netarhat which is always a great source of inspiration and knowledge to me.

Lastly to all those who have contributed in any way to the path I traverse.

Chapter-1

Number System

Synopsis with illustrated examples

Rational Number:

A number is called a rational number, if it can be written in the form of p/q , where p and q are integers and q is not equal to zero.

Decimal expansion of a rational number is either terminating or non-terminating recurring, while the decimal expansion of an irrational number is non-terminating.

Sum of two rational numbers and product of two rational numbers is always a rational number. Like ½ + ¼ = ¾ is a rational number.

Irrational Number:

Numbers which can be positioned on a number line but whole decimal expansion is non-terminating are called irrational numbers. Like $\sqrt{3}$ is a number which has its solution as non-terminating decimal, so it is irrational number.

Sum of one rational and one irrational number is an irrational number.

$5 + \sqrt{3}$ is an irrational number.

Product of one rational and other irrational number is an irrational number. Like $5 \times \sqrt{3}$ is an irrational number.

Real numbers:

Combined set of rational and irrational numbers is known as real numbers. These numbers can be placed on number line.

Imaginary numbers:

Numbers which are not real are known as imaginary numbers.

Digits:

1,2,3,4, 5, 6, 7,8, 9 and 0 are digits in the decimal system. Numbers are formed out of these digits only. Digits are also numbers of single digit. Examples of double digit numbers are 11, 25, 46, etc. Accordingly numbers with any number of digits can be formed.

Values of digits:

Digits have two type of values- one is real value of digit and the other is place value of digit.

Real value of digit is same as digit. Like in the number 125431, real value of digit 1 at rightmost and leftmost position is same, i.e., 1.

Place value of digit depends on the place at which the digit is there in the number. Place from rights are named as place of ones, tens, hundreds, thousands, ten thousands, lakhs, ten lakhs and so on. E.g.: in the number 125431, 1 at the rightmost place is at ones place and its place value is $1 \times 1 = 1$

But for left most digit 1, which is placed at place of lakhs, is $1 \times 100000 = 100000$.

So real value of digit remain the same and place value depends on the position of the digit in the number.

Indian system for naming of numbers or place value of numbers in decimal system:

The rightmost three digits are called ones:

Ones-

The rightmost is place of unit or one

Second from right is place of tens

Third from right is place of hundreds

Thousand-

Thousand- 4th from right

Ten thousand- 5th from right

Lakh-

Lakh- 6th from right

Ten Lakh- 7th from right

Crore-

Crore- 8th from right

Ten crore- 9th from right

After it next digits are for Arab, Ten Arab, Kharab, Ten Kharab.

In Table form we may show the same as follows:

Kharabs	Arabs	Crore	Lakh	Thousands	Ones

Ten Kharab	Kharab	Ten Arab	Arab	Ten Crore	Crore	Ten Lakh	Lakh	Ten thousand	Thousand	Hundreds	Tens	one

Natural Numbers:

As we know, we use 1, 2, 3, 4,... when we begin to count. They come naturally when we start counting. Hence, mathematicians call the counting numbers as Natural numbers.

So, Natural numbers are numbers starting with 1 and next natural number is 1 added to the previous number. Total infinite number of natural numbers are there.

Predecessor and successor

Given any natural number, you can add 1 to that number and get the next number i.e. you get its successor. Like successor to 301 is (301+1), i.e., 302.

The number 301 comes before 302, so 301, which is 302-1, is the predecessor of 302.

Natural number 1 has no natural number as its predecessor. Every natural number has one successor.

Successor of a natural number = Natural number + 1

Predecessor of a natural number = natural number -1

Whole Numbers:

The natural numbers along with zero form the collection of whole numbers.

So, whole number is a group of numbers in which all natural numbers are there and in addition zero is also there.

So, whole numbers are 0,1,2,3,4...... up to infinity

The Number Line

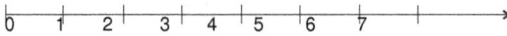

A line on which numbers are shown for their relative position. Such a line is obtained by fixing 0 & 1 and the using this distance as unit distance for showing other numbers. Number line shown above is for whole numbers.

On number line numbers shown on right are greater. So, successors are on right and predecessors on the left.

Addition and subtraction on the number line are possible. E.g. to add 3 to 2 we go three places right from 2, i.e. we reach at 5, so 3+ 2 = 5.

To subtract 2 from 3, we go 2 places left to 3, so that we get 1.

Multiplication on number line can also be performed accordingly by repeating the same distance on number line number of times it is to be multiplied with.

Exercise:

1. write the three next natural numbers after 100.

    Soln: first successor= 100 + 1= 101

    Number after 101 = 101 + 1 = 102

    Number after 102 = 102 + 1 = 103

    So, numbers are 101, 102 & 103.

2. which is the smallest whole number?

    Soln: As whole numbers start with 0, the smallest whole number is 0.

3. Write the successor of 30451.

    Soln: Successor of 30451 = 30451 + 1= 30452

4. Write the predecessor of 13245.

    Soln: Predecessor of 13245 = 13245-1= 13244

5. Which one is greater number: 97623 or 96521

    Soln: At place of thousand 7 is there in 97623 and 6 is there in 96521, so 97623 comes after 96521 in counting, so 97623 is greater. For comparison start comparing the digits from left most position and the number with greater left most position is greater.

6. Which one of the following is/ are true:
    a. Zero is the smallest natural number
    b. 402 is successor of 401
    c. 399 is predecessor of 398
    d. Zero is the smallest whole number
    e. All natural numbers are whole numbers.

    Soln:

    a. Zero is not a natural number but it is a whole number, so a. is incorrect and d is correct.
    b. 402= 401 + 1, so it is successor of 401.
    c. Predecessor of 398 = 398-1 = 397, so not correct.
    d. Correct
    e. Whole number is a group consisting of natural numbers with zero added to it.

Closure property of whole numbers under addition and multiplication:

Whole numbers are closed under addition and also under multiplication. It means that sum of any two whole numbers is always a whole number and multiplication of any two whole numbers is always a whole number.

Like 1+8=9, the 9 is also a whole number.

5 x 103 = 315, 315 is also a whole number.

But whole numbers are not closed under subtraction and division.

Like 1-3 = -2 is not a whole number. 1 ÷3 = 1/3 is not a whole number.

Commutativity of addition and multiplication of whole numbers:

Sum of any two whole numbers does not depend on order of addition of numbers.

So, **A + B = B + A** is known as commutative law of addition of whole numbers.

In the same way, **A x B = B x A**, is known as **commutative law** of multiplication for whole numbers.

Note that subtraction and division are not commutative for whole numbers. 1-3 is not equal to 3-1 and 1÷2 is not equal to 2÷1.

Associativity of addition and multiplication:

Grouping of numbers of addition does not change the result of addition.

i.e., **A + (B + C) = (A + B) + C** is known associativity of addition for whole numbers.

E.g: 10 + (20 + 30) = 10 + 50 = 60

Now changing the grouping, (10 + 20) + 30 = 30 + 30 = 60

So, associativity of addition holds true.

In the same way, **(A x B) x C = A x (B x C)** is known as associativity of multiplication of whole numbers.

Division by 0:

Division of whole number by zero, except zero shall be infinity, which is not a real number.

Division of 0 by 0 is not defined.

Exercise:

1.   123 + (201 + 11) = (103 +201)+ 11 is as per ....rule of addition.

Soln: It is as per rule of **associativity of addition** for whole numbers.

2.   560 x (12 x 30) = (560 x 12)x30 is as per ......rule or property.

Soln: It is as per rule of **associativity of multiplication** of whole numbers.

3. 12 + 435 = 435 + 12 is as per ......of addition.

Soln: Commutativity of addition for whole numbers.

3. Sum of two whole number shall be a while number is........property of whole numbers for addition.

Soln: It is as per closure property of whole numbers for addition.

Distributivity of multiplication over addition:

The property that A x (B+ C) = A x B + A x C, where A, B & C are whole numbers, is known as distributivity of multiplication over addition.

Exercise:

1.   Find 13 x 12

Soln: 13 X 12 = (10 +3) x 12

Now, (10+3) x 12 = 10 x 12 + 3 x 12 (This is as per distributivity of multiplication over addition)

= 120 + 36 = 156

2.   Find 133 x 23 + 67 x 23

Soln: 133 x 23 + 67 x 23 = 23 x 133 + 23 x 67 (Commutativity of multiplication)

= 23 (133 + 67) (distributivity of multiplication over addition)

= 23 (200) = 4600

Ans.: 4600

Identity of addition or additive identity:

By adding zero to any number we get the same number.

A + 0 = a

So, zero is known as identity of addition.

Identity of multiplication or multiplicative identity:

By multiplying any number with 1, we get the same number, so 1 is known as identity of multiplication. So, 1 is known as identity of multiplication.

Exercise:

1. Which of the following will not represent zero:

(a) 10 + 0 (b) 0 x 0 (c) 3 x 0 (d) 1÷0

Soln:

    (a)   10 + 0 = 10, Not equal to zero

    (b)   0 x 0 = 0

    (c)   3 x 0 =0

    (d)   1÷0 = infinity, which is not equal to zero

So, (a) & (d) are answer.

2.   Product of two numbers is zero, then one or both of them will be zero- Is it true?

Soln:

Say two numbers are x and y.

Now,$(x) \times (y) = o$,

So, if x is not zero, then $y = 0/ x = 0$

If, y is not zero, then $x = 0/ x = 0$

If x & y both are zero then $(x) \times (y)$ is zero.

So statement that if product of two numbers is zero, then one or both of them will be zero- is true.

3.   If the product of two whole numbers is 1, can we say that one or both of them will be 1? Justify through examples.

Soln: Let two numbers be X and Y.

Now, $X \times Y = 1$

So, $X = 1/Y$

If $y = 3$, then $x = 1/3$

So, the statement is not correct.

4.   Find using distributive property: 727 x 101

Soln: $727 \times 101 = 727 (100 + 1) = 72700 + 727 = 72000 + 700 + 727 = 72000 + 1427 = 73427$

5.   A taxi driver filled his car petrol tank with 30 litres of petrol on Monday. The next day, he filled the tank with 40 litres of petrol. If the petrol costs Rs 64 per litre, how much did he spend in all on petrol?

Soln: Total petrol bought on Monday and next day = 30 + 40 = 70 litre

Price of petrol = Rs 64 per liter

So, total spending on petrol = Rs. 64 x 70 = Rs. (60 + 4) x 70

= Rs. 60 x 70 + Rs. 4 x 70

= Rs. 4200 + Rs. 280

= Rs. 4480, Ans.

6. A vendor supplies 44 litres of milk to a hotel in the morning and 56 litres of milk in the evening. If the milk costs Rs 17 per litre, how much money is due to the vendor per day?

Soln:

Total milk given to the hotel per day = (44 + 56) Litre

So, money due to vendor per day = (44 + 56) Litre x Rs. 17 per litre

= 100 x Rs. 17 = Rs. 1700, Ans.

7. Match the following:

(a) (100 + 25) x 30 = 100 x 30 + 25 x 30	(i)	Commutative law
(b) 123 + 37 = 37 + 123	(ii)	Associative law of addition
(c) (123 + 45) + 34 = 123 + (45 + 34)	(iii)	Distributive law of multiplication over addition
(d) Identity of addition	(iv)	1
(e) Identity of multiplication	(v)	0

Soln:

(a)- (iii), (b)- (i), (c)- (ii), (d)- (v), (e)- (iv)

Summary

1. The numbers 1, 2, 3,... which we use for counting are known as natural numbers.

2. If you add 1 to a natural number, we get its successor. If you subtract 1 from a natural number, you get its predecessor.

3. Every natural number has a successor. Every natural number except 1 has a predecessor.

4. If we add the number zero to the collection of natural numbers, we get the collection of whole numbers. Thus, the numbers 0, 1, 2, 3,... form the collection of whole numbers.

5. Every whole number has a successor. Every whole number except zero has a predecessor.

6. All natural numbers are whole numbers, but all whole numbers are not natural numbers.

7. We take a line, mark a point on it and label it 0. We then mark out points to the right of 0, at equal intervals. Label them as 1, 2, 3,.... Thus, we have a number line with the whole numbers represented on it. We can easily perform the number operations of addition, subtraction and multiplication on the number line.

8. Addition corresponds to moving to the right on the number line, whereas subtraction corresponds to moving to the left. Multiplication corresponds to making jumps of equal distance starting from zero.

9. Adding two whole numbers always gives a whole number. Similarly, multiplying two whole numbers always gives a whole number. We say that whole numbers are closed under addition and also under multiplication. However, whole numbers are not closed under subtraction and under division.

11. Division by zero is not defined and is termed as infinity.

11. Zero is the identity for addition of whole numbers. The whole umber 1 is the identity for multiplication of whole numbers.

12. You can add two whole numbers in any order. You can multiply two whole numbers in any order. We say that addition and multiplication are commutative for whole numbers.

13. Addition and multiplication, both, are associative for whole numbers.

14. Multiplication is distributive over addition for whole numbers.

15. Commutativity, associativity and distributivity properties of whole numbers are useful in simplifying calculations and we use them without being aware of them.

Divisor, quotient and remainder:

If we devide 66 by 5.

5) 66 (13

```
     5
   --------
    16
    15
   ---------
     1
   ----------
```

If we see abovementioned devision, divisor is 5, whereas remainder is 1. Quotient is 13.

5 is not an exact divisor of 66, because 1 remains there as remainder.

If we take division of 65 by 5. Remainder is 0. So, 5 is exact divisor of 65.

65 can be written as, $65 = 5 \times 13$.

So a number is represented as a product of numbers, then each such number in the product side is known as a factor of number formed with multiplication.

E.g.: 5 and 13 are factors of 65 in this case.

We can understand that there are different combinations in which number can be written as product of other numbers and so more than two factors of a number are possible.

Any number can be represented as that number multiplied by 1, so that number and 1 are always factors of the number. Like for 65, 65 and 1 are factors.

So, we can write a number in the form of multiplication of exact divisors and get the factors of that number.

Q1: write all factors of 18.

Soln:

18 = 1x 18 = 1x2x9 = 1x2x3x3

So, prime factors of 18 are 2, 3 , 3

So, 1 & 18 are two factors as it is for every number.

Now 2x3x3 can be written with combination of multiplication on right hand side, it can be written as:

6 x 3 or 2x 9

So, all factors of 18 are 1,2,3,6,9 & 18.

How to make sure that no factor is left out:

Find prime factors of the number. Then write that number as a multiple with every prime factor. Include 1 and that number as they are factors in case of all the numbers. That way all factors can be listed out.

Like for 18, prime factors are 2, 3

So,

18 = 2 x 9 (factors 2 & 9)

18 = 3 x6 (factors 3 & 6)

Also that, 18 = 1x 18 (factors 1 & 18)

So, all possible factors of 18 are 1, 2, 3, 6, 9 & 18.

Note that 1 is not a prime number.

Some salient points of factors are as follows:

1. 1 is factor of every number.
2. The number itself is also a factor of that number.
3. Any factor of a number cannot be greater than that number or every factor is less than or equal to that number.
4. Number of factors of a given number are finite.

Multiples:

Any number is called to be a multiple of its factors. Like if A = B x C x D, A is multiple of B, C & D.

Like 4 is a multiple of 2, 4 = 2 x 2, 6 is a multiple of 2, 6 = 2 x 3.

We can also say that if some whole number a is multiplied by some other whole number b then result a.b is multiple of a.

Like, 2,4,6, 8, 10, all are multiples of 2.

Exercise:

1. Write four multiples of 21.

Soln:

21 x 1= 21

21 x 2= 42

21 x 3 = 63

21 x 4= 84

So, 21, 42, 63, 84 are four multiples of 21. But, many other multiples (in fact infinite) of 21 are possible.

Salient points about multiples are as follows:

1. Every multiple of a number is greater than or equal to that number
2. The number of multiples of a given number is infinite
3. Every number is a multiple of itself and is the smallest multiple.

Perfect Number:

A number for which sum of all its factors is equal to twice that number is known as perfect number.

For number 6, prime factors are 2 & 3

So,

6= 2 x 3

6 = 3 x2

6= 1 x6 (factors 1 and the number itself)

So, all factors of 6 are, 1, 2, 3 & 6.

Now, sum of all these factors is 1 + 2 + 3 + 6 = 12, which is twice (double) of 6. So, 6 is a perfect number. In the same way, it can be examined whether a given number is perfect number of not.

Exercise:

1. whether 68 is perfect number?

Soln:

Write 68 as a multiplication of prime numbers, by breaking and breaking numbers into factors till it is not possible to further break into multiples.

So, 68 = 2 x 34 = 2 x 2 x 17

Now, 2 & 17 both are prime and can not be further broken into factors.

So, prime factors of 68 are 2 & 17.

So, factors of 68 are:

1: for all the numbers

68: every number is a factor of itself

68/ 2 = 34

68/ 17 = 4

So, all factors of 68 are, 1, 2, 4, 17, 34 & 68.

Now sum of all factors of 68 = 1 + 2 + 4 + 17 + 34 + 68 = 126

Double of 68 is 68 x 2 = 136

As sum of all factors is not equal to twice the number, 68 is not a perfect number.

Prime and composite numbers:

A number which has only two factors (1 and itself) is known as prime number. That means a prime number is completely divisible by no number other than 1 & the number itself.

Some prime numbers are as listed in the below-mentioned table:

Number	Factors
2	1, 2
3	1,3
5	1,5
7	1,7
11	1, 11

1 is having only one factor 1 and no 2nd factor, so it is not a prime number.

Numbers having more than two factors are called **composite numbers.**

1 is neither prime nor composite number.

Twin Primes:

Any two prime numbers, whose difference is 2 called twin prime numbers or twin primes.

Like 3 & 5, 5 & 7, 11 & 13 are examples of twin primes.

Exercise:

1. Is 15 a prime or composite.

Soln:

15 = 3 x5

So, 15 has factors, 3 & 5 other than 1 & 15. So, it is having four factors (more than two) and hence a composite number.

Method given by Greek Mathematician **Eratosthenes** for finding prime numbers, called **sieve of Eratosthenes**:

Finding prime in the range 1 to any number:

Steps are :

Mark cross sign across numbers which are not prime and encircle numbers which are found to be prime.

Step 1: Cross out 1 because it is not a prime number.

Step 2 : Encircle 2, cross out all the multiples of 2, other than 2 itself, i.e. 4, 6, 8, 10 and so on.

Step 3 : You will find that the next uncrossed number is 3. Encircle 3 and cross out all the multiples of 3, other than 3 itself, like 6, 9, 12 and so on.

Step 4 : The next uncrossed number is 5. Encircle 5 and cross out all the multiples of 5 other than 5 itself.

Step 5 : Continue this process till all the numbers in the list are either encircled or crossed out.

All encircled numbers are prime numbers and all crossed numbers other than 1 are composite numbers.

Let us find prime number from 1 to 20 by this method.

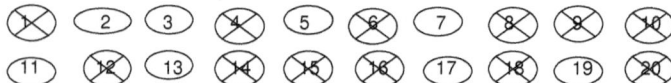

1 was crossed, as it is not prime.

Next number is 2, 2 is encircled as prime and multiples of 2 , i.e. 4, 6, 8, 10, 12, 14, 16 & 18 are crossed.

Next number is 3, which is not crossed, so its prime. Now it is encircled and its multiples, 6, 9, 12, 15 & 18 are crossed.

Next number 4 is already crossed. Next is 5, which is prime as it is not crossed.

Encircle 5 and cross multiples 10, 15 ad 20.

11 is prime and its multiple is 22, which is not in the range of 20.

In this way we found the prime numbers, which are encircled and are 2,3,5,7,11,13,17 and 19.

You may observe that after reaching midpoint of the range from 1, there is no need to see for multiples ay more because multiples of higher number shall lie outside the range now.

Even numbers:

Any number which is completely divisible by 2, i.e., with no remainder is known as even number.

So, all multiples of 2 like 2, 4, 6, 8, 10, 12 .. are even numbers.

In any multiple of 2, either of five digits is there at unit position- 0,2,4,6 or 8.

So, to check divisibility of any number by 2, see the ones place of the number if it is having 0,2,4,6 or 8 at ones place then it is divisible by 2 else it is not divisible by 2.

Smallest even number is 2, which is the only even prime number.

Odd number:

Numbers not divisible by 2 are called odd numbers.

Any number which is not an even number is an odd number. Numbers having 1, 3,5,7 & 9 at their ones place are not divisible by 2 and they are odd numbers.

Exercise:

1. Which of the following are even:
    a. 12345
    b. 2341
    c. 900823
    d. 0
    e. 3772

Soln: Ones place of 12345 is 5, so number not divisible by 2, so odd number.

For, 2341, at ones place 1 is there so not divisible by 2, so odd.

For 900823, at ones place 3 is there so odd.

O is neither even nor odd.

For 3772, at ones place 2 is there, so number is divisible by 2 and hence it is even number.

Ans.: e

2. Sum of one even and one odd numbers is always....number.

Soln:

Method1:

Say A is an even number and B is an odd number.

To check whether A + B is divisible by 2,

$(A + B)/2 = A/2 + B/2$

A is even number so A/2 is integer.

But, B is an odd number and so B/2 will not be with zero remainder, i.e., division will not be an integer.

So, (A +B)/ 2 will not be an integer, and hence it is not completely divisible by 2, so it is an odd number.

Method 2:

At ones place for even number, 2,4,6,8 or 0 shall be there.

For odd number at ones place, 1, 3, 5, 7 or 9 shall be there.

If we add any one number from first group with any number in the second group,

For e.g.: 2 + 1 = 3, 4 + 1 = 5 and so on

We note that in the resulting number at ones place, one out of 1,3,5,7 or 9 shall be there.

So, the resulting number shall be odd.

Method 3:

Even + Odd = Even + Even + 1= Even + 1= Odd number

So, odd number.

2. State whether the following statements are True or False:

(a) The sum of three odd numbers is even.

Soln:

Odd + Odd + Odd

 = Even + 1 + Even + 1 + Even + 1

= (Even + Even + Even) + 3

= Even + 2 + 1

= Even +1 = Odd

You can also check by example, 5 + 7 + 11 = 23 is an odd number. So **False.**

(b) The sum of two odd numbers and one even number is even.

Soln:

Odd + Odd + even = even + 1 + even + 1 + even

= (even + even + even) + 2

= even + 2 = even, **True**

(c) The product of three odd numbers is odd.

Soln:

Odd + Odd + Odd = Even + 1 + Even + 1 + Even + 1

= Even + 3

= Odd, so **True.**

(d) If an even number is divided by 2, the quotient is always odd.

Soln:

If even number is divided by 2 then it will have an integer as quotient and remainder will be zero.

The integer we get after division may or may not be divisible by 2.

e.g: 22÷2= 11 has quotient of 11, which is an odd number.

But, 24÷2= 12 has quotient of 12, which is an even number, So **False.**

(e) All prime numbers are odd.

Soln: False because 2 is one even number which is prime also.

(f) Prime numbers do not have any factors.

Soln: All numbers have at least two factors, one number itself and other 1, so statement is **False.**

(g) Sum of two prime numbers is always even.

Soln: Other than 2, all other prime numbers are odd. But if we add 2 (a prime number) to any other prime number then the sum will not be even, so statement is **false.**

(h) 2 is the only even prime number.

Soln: **True**

(i) All even numbers are composite numbers.

Soln: **False**, because 2 is one even and prime number

(j) The product of two even numbers is always even.

Soln: **True.**

3. Find twin primes less than 10.

Soln:

From 1 to 10, prime numbers are, 3, 5, 7. So, 3 & 5 are twin primes and 5 & 7 are also twin primes.

4. Which of the following numbers are prime? (a) 23   (b) 51   (c) 37   (d) 26

Soln: 23 is prime, as no factor other than 1 & 23.

51 = 3 x 17, so not prime

37 is also prime

26 = 2 x 13, so not prime

Thus, 23 and 37 are prime.

   (a) & (c) – Ans.

5. Express 31 as sum of three odd primes.

Soln: For this we need to know odd primes less than 31,

Listing out them, 3, 5, 7, 11, 13, 17, 19, 23 and 29

Start writing as sum with any prime number and check whether it is possible.

31 = 3 + 28 = 3 + 11 + 17 , Ans.

Try find other pairs,

Like 31 = 5 + 26

= 5 + 3 + 23

6. Fill in the blanks:

(a) A number which has only two factors is called a _____.

(b) A number which has more than two factors is called a _____. (c) 1 is neither _____ nor _____.

(d) The smallest prime number is _____.

(e) The smallest composite number is _____. (f) The smallest even number is _____.

Soln:

   (a) Prime
   (b) Composite number
   (c) 2
   (d) 4
   (e) 2

Tests for divisibility of numbers:

Divisibility by 2:

If we write few multiples of 2, we find them as 4, 6, 8, 10, 12, 14, 16….

Please note the pattern. Numbers divisible by 2 are having some pattern in the ones place of these numbers. These numbers have only the digits 0, 2, 4, 6, 8 in the ones place.

So, a number is divisible by 2 if it has any of the digits 0, 2, 4, 6 or 8 in ones place. Else the number is not divisible by 2.

<u>Divisibility by 3 :</u>

If we write multiples of 3, like 6, 9, 12, 15, 18, 21, 24, 27, 30, we see no pattern at its ones place.

Now, first two digit number formed with 3 is 12, whose sum of digits is (1+2) = 3, now if 3 is added to it to get next multiple, obviously next number 15 shall have 6 as sum of digit which is again divisible by 3, and so on.. so this is the pattern which can be used to check the divisibility by 3.

So, if the sum of the digits of a number is a multiple of 3, then the number is divisible by 3.

<u>Divisibility by 4 :</u>

100 is divisible by 4 and after that if we add 4, 8, 12, 16, etc. last digit shall remain the same and so these digits shall also be divisible by 4. Like 104, 112, 120, etc are divisible by 4.

The same thin is revisited for next hundreds. So, test of divisibility of 4 can be done from last two digits of the number.

A number with 3 or more digits is divisible by 4 if the number formed by its last two digits (i.e. ones and tens) is divisible by 4. Divisibility for 1 or 2 digit numbers by 4 has to be checked by actual division.

So, for checking division by 4, check whether the number formed by its last two digits (i.e. ones and tens) is divisible by 4. If so, number is divisible by 4.

Divisibility by 6 :

A number which is divisible by 6 must be divisible by its prime factors 2 and 3.

Let us take multiple of 6 = A, say

Then, A = 6 x n = (2 x 3) x n

= 2x 3n: multiple of 2

= 3 x 2n: multiple of 3

So, if a number is divisible by 2 and 3 both then it is divisible by 6 also.

So, to check divisibility by 6, check the divisibility of the number by 2 and 3 individually. If divisible by both then it is divisible by 6.

Divisibility by 8 :

1000 is divisible by 8, because 1000 = 125 x 8.

Now if 8, 16, 24, 32,......96, 104,.....992 are added to make next multiples. Now, these three digits are simply added multiples of 8 at last three digits, they are divisible by 8.

So, a number with 4 or more digits is divisible by 8, if the number formed by the last three digits is divisible by 8. The divisibility for numbers with 1, 2 or 3 digits by 8 has to be checked by actual division.

Divisibility by 9 :

If the sum of the digits of a number is divisible by 9, then the number itself is divisible by 9.

Divisibility by 11:

Thus, to check the divisibility of a number by 11, the rule is, **find the difference between the sum of the digits at odd places (from the right) and the sum of the digits at even places (from the right) of the number. If the difference is either 0 or divisible by 11, then the number is divisible by 11.**

Exercise:

1. Using divisibility tests, determine which of the following numbers are divisible by 4; by 8
    a) 432
    b) 45672
    c) 12004
    d) 1400
    e) 657891121
    f) 984352

Soln:

Divisibility by 4:

Number formed with ones and tens in all but e is divisible by 4, like in a) 32 ÷4 = 8, so divisible by 4. With last two digits of e), 21 is formed, which is not divisible by 4, so the number shall also not be divisible by 4.

Divisibility by 8:

a), 432 is divisible by 8

b) number with last three digits = 672, which is divisible by 8, so number divisible by 8.

c) number with last three digits = 004, not divisible by 8, so number not divisible by 8.

d) number with last three digits = 400, which is divisible by 8, so number is divisible by 8.

e) number with last three digits = 121 not divisible by 8, so number not divisible by 8.

f) number 352 divisible by 8, so number divisible by 8

2. Using divisibility tests, determine which of following numbers are divisible by 6:

(a) 297156   (b) 1264

Soln:

(a) last digit is 6, so number is divisible by 2.

Sum of the digits of the number = (2 + 9 + 7 + 1 + 5 + 6) = 30, which is divisible by 3, so number divisible by 3.

Number is divisible by both 2 & 3 so it is divisible by 6.

(b) last digit is 4, so number is divisible by 2.

Sum of the digits is (1 + 2 + 6 + 4) = 13, which is not divisible by 3, so number is not divisible by 3.

As, number is not divisible by 3, it shall not be divisible by 6.

3. Using divisibility tests, determine which of the following numbers are divisible by 11:

(a) 5775   (b) 10835   (c) 10000001

Soln:

(a) Sum of odd places = 5+ 7 = 12

Sum of even places = 5 + 7= 12

Difference of sum of odd and even places = 12- 12 = 0

As the difference is zero, the number shall be divisible by 11.

(b) Sum of odd places = (5 + 8 + 1) = 14, Sum of even places = 3 + 0 = 3, difference = 14- 3 = 11, divisible by 11, so number shall be divisible by 11.

(c) Sum of odd places = 1, even places = 1, difference = 0, so divisible by 11.

5. Write the smallest digit and the greatest digit in the blank space of each of the following

numbers so that the number formed is divisible by 3 :   **57**

(a) __6724   (b) 4765 __ 2

(a) sum of given digits = 6 + 7 + 2 + 4 = 19, let missing digit be x, then 19 + x must be divisible by 3 only then the number shall be divisible by 3. Number just greater than 19 and divisible by 3 is 21, so x = 21-19 =2. Ans.

(b) Sum of given digits = (4+7+6+5+2) = 24

Let missing digit be x,

Then 24 + x must be divisible by 3.

Now, 24 itself is divisible by 3, so the least possible digit at this place is 0.

6. Write a digit in the blank space of each of the following numbers so that the number formed is divisible by 11 :

(a) 92 __ 398   (b) 9 __ 9485

Soln:

(a)

Digits at odd places (from the right) are 8, 3, 2

Sum of these digits = 8 + 3 + 2 = 13

Let missing digit be x.

Now Digits at even places from right are 9, x, 9

So, sum of digits at even places (from the right) = 9 + x + 9 = 18 + x

Sum of digits at even places – Sum of digits at odd places = 18 + x – 13 = 5 + x

So, 5 + x , should be either 0 or divisible by 11.

So, 5 + x = 11

Or, x = 6. Ans.

(b) Let missing digit = x

So, digits at odd places (from the right) = 5, 4, x

So, sum of these digits at odd places = 5 + 4 + x = 9+ x

Digits at even places = 8, 9, 9

So sum of digits at even places = 8 + 9 + 9 = 26

So, difference = 26 –(9+ x) = 17 – x

So, 17 – x = either 0 or multiple of 11

If, 17 – x = o, then x = 17, which is not possible, because it is a single digit number

So, 17 – x = 11

Or, x = 17 -11 =6, Ans.

Some more rules of divisibility:

1. **If a number is divisible by another number, then it is divisible by each of the factors of that number.**

For example, 420 is divisible by 42 and 6 is a factor of 42, then 420 is divisible by 6 also and by any other factor of 42, like 3, 7 and 21.

2. **If a number is divisible by two co-prime numbers (whose highest common factor is 1) then it is divisible by their product also.**

For example, 120 is divisible by co-prime nos. 6 and 5, so120 is divisible by 6 x 5 = 30.

3. **If two given numbers are divisible by a number, then their sum is also divisible by that number.**

For example, 60 is divisible by 3 and 12 is divisible by 3 so, (60 + 12), i.e., 72 is also divisible by 3.

4. **If two given numbers are divisible by a number, then their difference is also divisible by that number.**

For example, 60 is divisible by 3 and 12 is divisible by 3, then (60 – 12), i.e., 48 is divisible by 3.

Roman Number System

The number system we use in this course is based on the Hindu-Arabic system which uses the digits 0, 1, 2, 3, 4, 5, 6, 7, 8 and 9. Romans used to use a different number system, known as Roman system. Symbols corresponding to Hindu Arabic system used in Roman System are as follows:

Hindu Arabic system	1	2	3	4	5	6	7	8	9	10	20	30	40	50	60	100	500	1000
Roman system	I	II	III	IV	V	VI	VII	VIII	IX	X	XX	XXX	XXXX	L	LX	C	D	M

Numbers written in the Roman system had to be written in order.

Small number symbol written before (on left of) large number symbol means net number = large number – small number

For example:

IV stands for 1 before 5   or 4

XC stands for 10 before 100   or 90

Small number symbol written after (on right of) large number symbol means net number = large number + small number

For example,

VI stands for 1 after 5 or 1+ 5 = 6

CX stands for 10 after 100 or 110

Noticeable rules regarding which possess restriction of which symbols can appear before are:

- I could only appear before V or X.
- X could only appear before L or C.
- C could only appear before D or M.

One less than a thousand was therefore not written as IM but as CMXCIX.

Large numerals were formed by placing a stroke above the symbol. A stroke above a numeral makes the numeral 1000 time.

For example,

$\overline{V} = 5 \times 1000 = 5000$

$\overline{X} = 10 \times 1000 = 10000$

$\overline{L} = 50 \times 1000 = 50000$

Exercise:

1. What numbers are represented by the following symbols?

a. VIII	b. XII	c. XIV	d. X$\overline{VI}$	e. XXXI
b. LXXXI	f. MCLVI	g DCCXX	h. $\overline{LX}$	

Soln:

    a. VIII = 5 + 1 + 1 + 1 = 8
    b. XII = 10 + 2 = 12
    c. XIV = 10 + (5-1) = 14
    d. XVI = 10 + 5 + 1 = 16
    e. XXXI = 10 x 3 + 1 = 31
    f. LXXXI = 50 + (3 x 10) + 1 = 81
    g. MCLVI = 1000 + 100 + 50 + 5 + 1 = 1156
    h. DCCXX= 500 + 100 + 100 + 10 + 10 = 720
    i. $\overline{LX}$ = 1000 x (50 + 10) = 60000

2. Write the following numbers in Roman numerals:

   a 19   b 34   c 289   d 903   e 1048   f 2553

Soln:

    a. 19 = 10 + 10 -1 = IXXX
    b. 34 = 10 +10 + 10 + 4 = XXXIV
    c. **289 = 100 + 100 + 100 – 11 = XICCC**
    d. **903 = 1000 – 97 = 1000 –(100 -3)= IIICM**

3. Write the year 1999 using Roman symbols

Soln:

1999= 2000 – 1 = -100 + (2000) + 99= CMMXCIX.

Note that it is not written like IMM, because I is written only on left side of V or X.

Chapter-2

Solved Examples

Q1: Write 789345621 in words

Soln:

Writing so that place multipliers are written just above the digits:

Crore		Lakh		Thousands		Ones		
Ten Crore	Crore	Ten lakh	One lakh	Ten thousand	Thousand	Hundreds	Tens	One
7	8	9	3	4	5	6	2	1
Seventy eight crore		Ninety three lakh		Forty five thousand		Six hundred twenty one		

In the third row, digits of the number are arranged from left to right. For writing in words, we write numbers in separate positions like ones, thousands, lakh and crore one by one and club together to write the number.

So, number becomes, **seventy eight crore ninety three lakh forty five thousand six hundred twenty one.**

Ans.

Q2: Write 308506008230 in words.

Soln:

Writing in place value table,

Kharab		Arab		Crore		Lakh		Thousands		Ones		
Ten Kharab	Kharab	Ten Arab	Arab	Ten Crore	Crore	Ten lakh	One lakh	Ten thousand	Thousand	Hundreds	Tens	One
	3	0	8	5	0	6	0	0	8	2	3	0
	Three	Eight		Fifty		Sixty		Eight		Two hundred thirty		

As mentioned in the lowest row of the table,

The number is words is,

Three Kharab eight arab fifty crore sixty lakh eight thousand two hundred thirty.

Note: lakh, crore, arab, kharab, etc. are hindi numerations. International system of numbers is in million, billion, trillion, etc.

Q3: Write in number form: 15000050

Soln:

Arranging digits as per their place value in table.

Crore		Lakh		Thousands		Ones		
Ten Crore	Crore	Ten lakh	One lakh	Ten thousand	Thousand	Hundreds	Tens	One
	1	5	0	0	0	0	5	0
One		fifty		zero		fifty		

So, number in words is, one crore fifty lakh fifty.

Note; zero is there in thousands, so it is not mentioned in the word form.

Q4: two students were asked to write seven thousand seven hundred seven. First student wrote it as 7000707 and second as 77007. Tell the error committed by them.

Soln:

First student has first written seven thousand and then wrote seven hundred seven followed by that. In this way numbers are not placed at right places and he wrote seventy lacs seven hundred seven.

Second student wrote 7700 first and then followed by that wrote 7, which is again not the correct way. In this way he wrote seventy seven thousand seven.

The write way shall be placing digits at write place in place value table, as shown below:

Thousands		Ones		
Ten thousand	Thousand	Hundreds	Tens	One
	7	7	0	7
seven		Seven hundred seven		

Write a number in expanded form:

To write number in expanded form follow following steps:

- Write number as addition of numbers with each digit followed by all places on the right of the digit replaced with zero.
- Now, if three zero are there on the right side of a number, then multiplying factor with that number shall be 1000, i.e. $10 \times 10 \times 10 = 10^3$. So, n number of zero on right means multiplication of that digit by 10^n.
- For example, for number ABCDE, with digits ABCDE, we can write in expanded form like this:
  - On right side of A, four digits B, C, D and E are there, so in place of A, we shall write, A0000, i.e., $A \times 10^4$
  - On right side of B, three digits are there, so in place of B we will write B000, i.e., $B \times 10^3$
  - For C, C00
  - For D, D0
  - For E, E0
- So, the number can be written like this in expanded form, $A \times 10^4 + B \times 10^3 + C \times 10^3 + D \times 10^2 + E \times 10^1$
- For number with higher number of places shall also be expanded in same way.
- So, if a digit is at nth place from right, power of 10 shall be n-1.

Q5: write 3251 in powers of ten.

Soln:

Method-1:

3251 = 3000 + 200 + 50 + 1

$= 3 \times 10^3 + 2 \times 10^2 + 5 \times 10^1 + 1 \times 10^0$

Ans.

Method-2:

Directly using (n-1) power for 10 for nth place from right:

$3 \times 10^{(4-1)} + 2 \times 10^{(3-1)} + 5 \times 10^{(2-1)} + 1 \times 10^{(1-1)}$

$= 3 \times 10^3 + 2 \times 10^2 + 5 \times 10^1 + 1 \times 10^0$, **Ans.**

Q6: Write the number which represents the expanded form: $2 \times 10^3 + 3 \times 10^2 + 0 \times 10^1 + 5$

Soln:

Method1:

$2 \times 10^3 + 3 \times 10^2 + 0 \times 10^1 + 5$

$= 2 \times 1000 + 3 \times 100 + 0 + 5$

$= 2000 + 300 + 5 = 2305$, Ans.

Method2:

Rule: 10^n is placed at (n+1) th position from right.

$2 \times 10^3 + 3 \times 10^2 + 0 \times 10^1 + 5 \times 10^0$

2 at (3+1), i.e. 4th place from right

3 at (2+ 1) i.e. 3rd place from right

0 at (1+ 1) i.e. 2nd place from right

5 at (0+ 1), i.e 1st position from right

4th	3rd	2nd	1st
10^3	10^2	10^1	10^0
2	3	0	5

So, the number is 2305, **Ans.**

Note: 10^0 is 1. In fact any number raised to power zero is 1.

Place value of digits:

Place value of a digit in a number is dependent on the place at which it is there in the number. For example:

- A digit at place of ones has place value of digit x 1 (first from right)
- A digit at place of tens has place value of digit x 10 (second from right)
- A digit at place of hundred has place value of digit x 100 (Third from right)
- A digit at place of thousand has place value of digit x 1000 (Fourth from right)
- So, a digit at nth place from right has a place value of digit x 10^{n-1}

True value of digit:

True value of digit is equal to the digit itself, because it is the value which does not change with respect to the position of digit in the number.

Like in 1311, all 1 digits have the same true value of 1 and 3 has true value of 3.

Q7: What is the place value of digits in the following:

(a) 359:

Soln: 9 at place of ones, place value = $9 \times 10^0 = 9 \times 1 = 9$

5 at place of tens, place value of 5 = $5 \times 10 = 50$

3 at place value of hundreds, place value of 3 = $3 \times 10^2 = 3 \times 100 = 300$

Alternate method:

To get place value of a digit, simply put equal number of zero as it is there in the number, after that digit .

For example,

359:

9	9
5	50 (one zero in place of one digit on right of 5)
3	300 (two zero in place of two digits on right of 3- 5 & 9)

(b) 4203:

Soln:

Digit	Place value	Remarks
3	3	3 is the rightmost digit, so no zero after 3.
0	00 = 0	One digit on right side, but digit itself is zero, so place value is zero.
2	200	Two digits (0 &3) on right os 2, so 2 followed by two zero
4	4000	Three digits (203) on the right so place value is 4000.

(c) **70809**

Soln:

Digit	Place value	Remarks
7	7 x 10000= 70000	5th place of right- place of ten thousands
0	0 x 1000= 0	4th place from right- place of thousands
8	8 x 100= 800	3rd place from right- place of hundreds
0	0 x 10= 0	2nd place from right- place of tens
9	9 x 1= 9	1st place from right

Largest numbers formed with digits:

Q8: Write the largest and the smallest numbers of seven digits.

Soln: For seven digit number, there are seven places, which can be filled with any digit from 0 to 9, except the first digit from left, cannot be zero.

To have the largest number one should place the largest digit at the highest place value.

For a nine digit number:

Place number	9th place	8th place	7th place	6th place	5th place	4th place	3rd place	2nd place	1st place
Which number can be placed in general	Any digit from 1 to 9, but 0 can not be placed, because then number will not be of nine digits	Any digit from 0 to 9	Any digit from 0 to 9	Any digit from 0 to 9	Any digit from 0 to 9	Any digit from 0 to 9	Any digit from 0 to 9	Any digit from 0 to 9	Any digit from 0 to 9
For number to be the highest nine digit number	9 (The highest possible digit)	9	9	9	9	9	9	9	9
For number to be the lowest number of nine digit	1 (The lowest possible digit)	0 (The lowest possible digit)	0	0	0	0	0	0	0 (The lowest possible digit)

So, largest number of nine digit = 999999999, Ans.

Note:

- To get highest number of any number of digits, put all 9 in the number, if no limitation of available digits is give.
- To get lowest number of any number of digits, put 1 at left most place and put 0 at other places.

Q8: Smallest number of seven digit.

Ans.:

As explained earlier, for smallest number 1 at the left most place and other places shall be occupied with zero.

So, the smallest number of seven digits = 1000000, **Ans.**

Q9: Find sum of the largest number of four digits and the smallest number of four digits.

Soln:

The largest number of four digits = 9999

The smallest number of four digits = 1000

So, Sum = 9999 + 1000 = 10999, **Ans.**

Q10: What will be the remainder when the largest number of four digits is divided by 10.

Soln:

The largest number of four digits = 9999

To find remainder when 9999 is divided by 10.

9999 = 9990 + 9

9990 is divisible by 10 because it is having 0 at ones place (rightmost) and 9 is less than 10. So, remainder in this case is 9.

Ans.: 9 (You can check through dividing also).

Q11: How many numbers of five digits are there?

Soln:

For finding total numbers of five digit numbers, we should know the highest four digit number (because after that the first number shall the lowest five digit number) and the highest five digit number and the get the difference of these two.

The Highest four digit number = 9999

The highest five digit number = 99999

So, total number of five digits numbers = 99999- 9999 = 90000, Ans.

Note : In the same way, total number of numbers of n digits will be 9(followed by n-1) zero.

Total number of:

- One digit numbers = 9
- Two digit numbers = 90
- Three digit numbers = 900
- Four digit numbers = 9000,
- Ten digit numbers = 9000000000

Largest and smallest digits made out of given digits:

Largest:

- Fix number of places equal to number of digits
- At the leftmost place, place the largest digit.
- Move to the right by putting smaller numbers

Smallest:

- Fix number of places equal to number of digits
- Keep the lowest digit other than 0 at the leftmost place
- At 2nd place put the lowest digit left among the group
- Move places to the right and place the lowest among remaining and so on up to the rightmost place.

Q12: Find difference of the highest number and the lowest number of three digits formed with 7, 0 & 3.

Soln:

Given digits are 7, 0 and 3.

For the highest number:

The highest digit	The next lower digit	The lowest digit
7	3	0

Thus, the highest number formed is 730.

For the lowest number:

The lowest digit other than 0	The lowest among remaining	The lowest among remaining
3	0	7

Thus, the lowest number formed is 307.

So, the difference = 730 – 307 = 423, **Ans.**

Q13:

How many natural numbers less than 100 are divisible by 3?

Soln:

For this, let us see the trend. 1st such number is 3, followed by 6, 9, 12.. with an increase of 3. The last number up to 100, which shall be divisible by 3 is 99.

3 =1 x 3 (1st number)

6 = 2 x 3 (2nd number)

..

99 = 33 x 3 (33rd number)

So, total number of numbers less than 100, divisible by 3 is 33.

Ans.: 33

Q14: How many numbers are there up to 200, which are divisible by both 3 and 7.

Soln:

3 and 7 are prime numbers, so a number which is divisible by 3 and 7 shall be divisible by 3 x 7, i.e. 21.

So, we have to find no of numbers up to 200, which are divisible by 21.

Dividing, 200/ 21 = 9, with remainder of 11.

So, total 9 numbers are there before 200, which are divisible by 3 and 7. They are, 21 x 1, 21 x 2, 21 x 2, ...21 x 9.

Ans.: 9

Q15: How many numbers are there up to 200, which are either divisible by 3 or 7.

Soln:

Number of numbers which are divisible by 3 or 7 = (Number of numbers divisible by 3) + (Number of numbers divisible by 7) – (Number of numbers divisible by both 3 & 7)

Subtraction is done to ensure than numbers are not counted two times, because some numbers like 21, are divisible by 3 and 7 both.

Number of numbers divisible by 3 = Quotient part of (200 ÷ 3) = 66

Number of numbers divisible by 7 = Quotient part of (200 ÷ 7) = 28

Numbers which are divisible by both 3 and 7:

These numbers shall be divisible by 3 x 7= 21, because 3 and 7 are prime numbers.

So, number of numbers divisible by 21 = Quotient part of (200 ÷ 21) = 9

So, number of numbers which are divisible by 3 or 7 = 66 + 28 − 9 = 85, **Ans.**

Q16: Find number between 400 and 500 which is completely divisible by 7 and 13.

Soln:

7 and13 are prime numbers, so a number which is divisible by 7 and 13 shall be divisible by 7 x 13 = 91.

Quotient of (500 ÷ 91) = 5 (Divide and see)

So, number divisible by 91 = 91 x 5 = 455, which is between 400 and 500.

Ans.: 455

Q17: Find sum of natural numbers from 1 to 20.

Soln:

Formulae: Sum of natural numbers from 1 to n = (n) (n+ 1) / 2

Sum of natural numbers from 1 to 20 = (20) x (20 + 1)/ 2= 10 x 21 = 210, Ans.

Q18: Find the sum of even numbers between 1 and 30.

Soln:

Sum of even numbers between 1 and 30 = 2 + 4 + ...+ 28

= 2(2/2 + 4/2 + 6/2+.....+ 28/2)

= 2(1 + 2 + 3+ .....14)

= 2(14 x (14+1))/2= 14 x 15 = 210, **Ans.**

Q19: Find the sum of all numbers in the table of 3.

Soln:

Sum = 3 + 6 + 9 + 12 +........+ 30

= 3(1 + 2+ 3+......+ 10)

= 3 (10) x (10+1)/ 2

= 3 x 10/2 x 11 = 165, **Ans.**

Q20: Find the sum of all the numbers in the table of 20.

Soln:

Sum = 20 + 40 + 60 +........+ 200

= 20 (1 + 2 + 3+.......+ 10)

= 20 (10 x 11)/2 = 10 x 10 x 11 = 110, **Ans.**

Q21: Find the sum of numbers from 10 to 20.

Soln:

Sum = 10 + 11 + ...+ 20

= (1+ 2 + 3 + ...+ 10)- (1+ 2 + 3 + ...+ 10) + (10 + 11 +.....+ 20)

=(1+ 2 + 3 + ...+ 10) + (10 + 11 +.....+ 20)- (1+ 2 + 3 + ...+ 10)

= (1+ 2 + 3+....+ 20) − (1 + 2 + 3+ .....+10)

= (20 x 21)/2 − (10 x 11)/2

= 210 − 55 = 155, **Ans.**

Q22: Eretosthense method : Find composite number from 1 to 100.

Soln:

Steps are :

Mark cross sign across numbers which are not prime and encircle numbers which are found to be prime.

Step1: Cross out 1 because it is not a prime number.

Step 2 : Encircle 2, cross out all the multiples of 2, other than 2 itself, i.e. 4, 6, 8, 10 and so on.

Step 3 : You will find that the next uncrossed number is 3. Encircle 3 and cross out all the multiples of 3, other than 3 itself, like 6, 9, 12 and so on.

Step 4 : The next uncrossed number is 5. Encircle 5 and cross out all the multiples of 5 other than 5 itself.

Step 5 : Continue this process till all the numbers in the list are either encircled or crossed out.

All encircled numbers are prime numbers and all crossed numbers other than 1 are composite numbers.

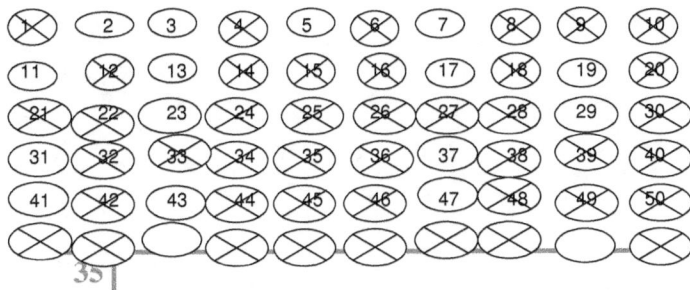

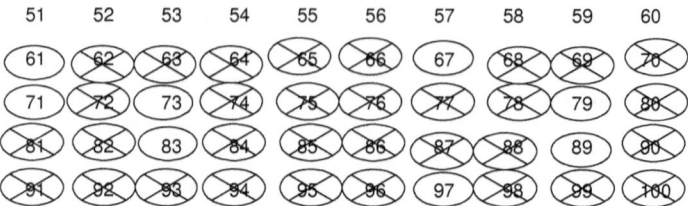

| 51 | 52 | 53 | 54 | 55 | 56 | 57 | 58 | 59 | 60 |

Following the steps mentioned before, prime numbers are encircled and composite numbers are crossed. So, all numbers except following are composite numbers:

1, 2, 3, 5, 7, 11, 13, 17, 19, 23, 29, 31, 41, 43, 47, 53, 59, 61, 67, 73, 79, 83, 89, 97. **Ans.**

Problems on number systems on different bases, conversion from one base to other base:

For number at base n,

Nth			4th	3rd	2nd	1st
Digit x n^{N-1}			Digit x n^{4-1}	Digit x n^{3-1}	Digit x n^{2-1}	Digit x n^{0}

For number at base N, numbers up to N-1 are used. Like for number at base 12, numbers up to 11 (with symbol e) is used.

Q23: $(4231)_5 = (......)_{10}$

Soln:

$(4231)_5 = 1 \times 5^0 + 3 \times 5^1 + 2 \times 5^2 + 4 \times 5^3$

$= 1 \times 1 + 3 \times 5 + 2 \times 25 + 4 \times 125$

$= 1 + 15 + 50 + 500 = 566,$

$(566)_{10},$ **Ans.**

Q24: Whether $(42)_5 = (22)_{10}$

Soln:

$(42)_5 = 2 \times 5^0 + 4 \times 5^1$

$= 2 \times 1 + 4 \times 5 = 22$

$= (22)_{10},$ **Ans. True**

Q25: Which number is equal to $(2te)_{12}$

Soln:

(t) is ten and (e) is eleven when number is on base of twelve.

$(2te)_{12} = 11 \times 12^0 + 10 \times 12 + 2 \times 12 \times 12 = 11 + 120 + 288 = (419)_{10}$, **Ans.**

Q26: convert $(1011)_2$ to decimal system.

Soln:

$(1011)_2 = 1 \times 1 + 1 \times 2 + 0 \times 2 \times 2 + 1 \times 2 \times 2 \times 2 = 11 = (11)_{10}$, **Ans.**

Objective Questions:

1. How may prime numbers are there less than 50:
(i)   50   (ii) 32   (iii) 16   (iv) 15
2. Which one of the following is the highest number, which is divisible by 2 and 8:
(i)   0   (ii) 8   (iii) 12   (iv) 16
3. Which one of the following is prime number:
(i) $\sqrt{93}$ (ii) $\sqrt{\frac{2}{7}}$ (ii) $\sqrt{\frac{9}{10}}$ (iv) $\sqrt{\frac{3}{27}}$
4. Which one of the following is a prime umber:
(i) $\sqrt{\frac{4}{3}}$ (ii) $\sqrt{\frac{0}{5}}$ (iii) $\sqrt{\frac{6}{8}}$ (iv) $9/\sqrt{16}$
5. The smallest natural number is:
(i)   1   (ii) 2   (iii) 3   (iv) 0
6. The smallest integer out of the following is:
(i)   1   (ii) 2   (iii) -1   (iv) 0
7. The smallest integer is:
(i)   0   (ii) 1   (iii) +/- 10   (iv) -1
8. Prime number between 90 and 100 is:
(i)   91   (ii) 93   (iii) 97   (iv) 99
9. Which of the following Is not positive integer:
(i)   8 + 4   (ii) $3\sqrt{2}$ (iii) (-4) x (-3)   (iv) (-8) ÷ (-4)
10. Example of complete series of even integers is :
(i)   2, 4, 6, 8   (ii) 1, 3, 5, 7   (iii) +/- 2, +/- 4, +/- 6, +/- 8   (iv) +/1, +/- 3, +/- 5, +/- 7
11. Real number is said to be:
(i)   Only rational number   (ii) Only irrational number   (iii) combined set of rational and
       irrational number   (iv) Neither rational nor irrational number
12. Tick (√) in front of correct statements and (x) against wrong statements:
    (i)   137532 is not divisible by 4   ( )
    (ii)   4, 6, 8, 9, 10 are prime numbers   ( )
    (iii)   2 is a composite number   ( )
    (iv)   15 is a prime number   ( )
    (v) $(42)_7 = (30)_{10}$ ( )
    (vi) $(23)_5 = (31)_4$ ( )
    (vii) $(24)_5 = (15)_{10}$ ( )
    (viii) $(111)_2 = (7)_{15}$ ( )

Other Problems

13. Fill in the blanks after understanding the pattern:
(i)   2, 5, 10, 17, 26.......
(ii)   8, 12, 9, 16, 10, 20, 11...........
(iii)   8, 24, 27, 9, 27, 30, 10........
(iv)   3, 6, 11, 18, 27, ...
(v)   2, 6, 12, 20, 30, ...

(vi)	1, 4, 7, 10, 13,.......		
(vii)	53, 37, 27, 17,........		
(viii)	2, 5, 8, 11, ...		
(ix)	3, -7, -11, -15, -19,........		
(x)	2, 9, 28, 71,......		
(xi)	42, 37, 32, 27,......		

(xii)	7	14	12
	4	12	9
	6	24	....

(xiii)	30	17	21
	18	15	27
	6	.....	30

14. Fill in the blanks below:

(i)   -5 is ....than -1

(ii)   (3+ 5) + 7 = 3 + (5 +7) explains the ....law of integers.

(iii)   If a and b are prime numbers then a + b is also a prime number. This law is known as ....law.- check

(iv)   Additive identity for positive integers is ....

15. Write answer of following in front of questions:

(i)   Sum of two inegers is an integer is according to which property of inegers....................

(ii)   For rational numbers, multiplicative identity is...................

(iii)   A (B + C) = AB + AC is as per....................law

(iv)   A x (B x C) = (A x B) x C, shows ....property of numbers.

16. 7 x (3 x 8) = (7 x 3) x 8, is based on ....law of product of numbers.

17. 7 + (-13) = (-13) + 7, is as per ....law.

18. Which is greater number, -4/5 or -7/9.

19. Express 1000 in powers of 10.

20. What is sign of number got through raising negative power to odd number.

21. What is exponent (power), if 64 is written in powers of 2.

22. a(b+c) = ab + ac, shows........................law

23. which one is greater: 3/5 or 4/5

24. which one is greater: 7/8 or 2/3

25. $\{(5)^2\}^3$ is equal to:

        (i)   5 x 2 x 3   (ii) 25 x 3   (iii) $(5)^5$ (iv) $(5)^6$

26. ....multiplicative identity for natural numbers.

27. Which numbers are represented by following:

        (i) $(2t2)_{12}$ (ii) $(5te)_{12}$ (iii) $(e5t)_{12}$

28. In which system only (0, 1, 2) digits are used.

29. Write first four natural numbers.

30. Write first four even numbers.

31. Write first four odd numbers.

32. Write one prime number between 5 and 10.

33. Express 0.12 as rational number.

34. What is additive inverse of 5.

35. What is multiplicative inverse of -1/4.
36. Which number is highest two digit prime number less than 42.
37. Which number is with three
38. (4 + 13) = (13 + 4) is as per..........................law of addition of integers.
39. Find the largest number of three digits which is divisible by 5.
40. Which even number is a prime number also.
41. The smallest number of five digits is.....................
42. (LXVI) is equal to.........................
43. Value of $(3)^3$ is equal to ....
44. Additive inverse of 8/3 is ....
45. Value of $\{(5)^2\}^3$ is..........................
46. If 10000 is written as powers of 10, what is exponent (power).
47. The smallest prime number between 1 and 10 is ....
48. Write two odd prime numbers, whose sum is 32.

Exercise and Solutions

1. How may prime numbers are there less than 50:

Soln: Follow eratosthense method of finding prime numbers between 1 and 50. Prime numbers found are:

2, 3, 5, 7, 11, 13, 17, 19, 23, 29, 31, 37, 41, 43, 47, i.e., total 15 number of prime numbers.

Ans.: (iv) 15

2. Which one of the following is the highest number, which is divisible by 2 and 8:

Soln: Out of four numbers given (0, 8, 12 and 16), the highest number is 16 and it is divisible by 2 as well as by 8, so, answer is 16.

Ans.: (iv) 16

3. Which one of the following is rational number:
   (ii) $\sqrt{93}$: Square root value not firm, so irrational number, so can not be prime.

   (ii) $\sqrt{\frac{2}{7}}$: Irrational number, so not rational.

   (ii) $\sqrt{\frac{9}{10}}$ = 3/ $\sqrt{10}$ and square root of 10 cannot be expressed as firm up value, so irrational number, Hence, not a rational number.

   (iv) $\sqrt{\frac{3}{27}} = \sqrt{\frac{1}{9}}$ = 1/3, which is a rational number.

   Ans.: (iv)

4. Which one of the following is a rational umber:
   (i) $\sqrt{\frac{4}{3}}$ (ii) $\sqrt{\frac{0}{5}}$ (iii) $\sqrt{\frac{6}{8}}$ (iv) 9/ $\sqrt{16}$

Soln: For numbers under square root:

   A. Square root should result in either integer, or fraction with integers as numerator and denominators where denominator is non-zero. This is possible in case square roots result comes out as terminating decimal or repeating decimal.
   B. If result is indefinite decimal digits, which are neither ending nor having a pattern of repetition, then number shall be irrational number.

In this problem, (i) is $\sqrt{4}$ /$\sqrt{3}$ = 2 / $\sqrt{3}$, Now $\sqrt{3}$ is neither digit, nor terminating decimal, nor repeating decimal, so it is irrational number. So, this is an irrational number.

Option (ii) is $\sqrt{\frac{0}{5}}$, which becomes 0/ $\sqrt{5}$ = 0, which is rational number.

Option (iii) is $\sqrt{\frac{6}{8}} = \sqrt{\frac{3}{4}} = \sqrt{3}/ 2$, so irrational number.

41

Option (iv) is $9/\sqrt{16} = 9/4$, which is ratio of two integers and hence a prime number.

5. The smallest natural number is:
(i)   1   (ii) 2   (iii) 3   (iv) 0

Soln: Natural numbers start from 1. So, out of the given numbers, 1 is the lowest natural number. Ans. (i)

6. The smallest integer out of the following is:
(ii)   1   (ii) 2   (iii) -1   (iv) 0

Soln: Integers can be positive or negative including zero. So out of the given integers, -1 is the smallest integer. Ans (iii)

7. The smallest integer is:
(ii)   0   (ii) 1   (iii) +/- 10   (iv) -1

Soln: Integer, -10, which is part of +/- 10 is the smallest integer, so, Ans. (iii)

8. Prime number between 90 and 100 is:
(ii)   91   (ii) 93   (iii) 97   (iv) 99

Soln: Numbers between 90 and 100 are, 91, 92, 93, 94, 95, 96, 97, 98 & 99,

Even numbers in the group are not odd numbers, because they all will be divisible by 2, so on

so rest odd numbers are:

91, 93, 95, 97, 99

91 = 7 x 13

For 93, sum of digits = 9 + 3 = 12,which is divisible by 3, so it is divisible by 3. 93 = 3 x 31

95, has 5 at its ones place, so divisible by 5.

97 has no other factor than 1 and 97.

99 is divisible by 3, so not prime.

Thus, only prime number between 90 and 100 is 97.

Ans. (iii) 97

9. Which of the following Is not positive integer:
(i)   8 + 4   (ii) $3\sqrt{2}$ (iii) (-4) x (-3)   (iv) (-8) ÷ (-4)

Soln: 8 + 4 = 12, so positive integer, (-4) x (-3) = 12, (-8) ÷ (-4) = 2

But $3\sqrt{2}$ has $\sqrt{2}$ as a multiplication term and it is irrational because solving for square root, we get 1.414... and it is still unending without any recurring pattern. So, product of an irrational number with an integer shall be irrational number.

10. Example of complete series of even integers is :

(i)   2, 4, 6, 8   (ii) 1, 3, 5, 7   (iii) +/- 2, +/- 4, +/- 6, +/- 8   (iv) +/1, +/- 3, +/- 5, +/- 7

Soln: Integers can be positive as well as negative. So, complete series of even integers is (iii)

11. Real number is said to be:

(i)   Only rational number   (ii) Only irrational number   (iii) combined set of rational and
      irrational number   (iv) Neither rational nor irrational number

Soln: Real number is a group of numbers in which numbers are either rational or irrational. So, (iii) is correct.

12. Tick (√) in front of correct statements and (x) against wrong statements:

(i)   137532 is not divisible by 4   ( )

Soln: for divisibility by 4, check for the divisibility of rightmost two digits in the integer. For given number the digit 32 (number formed with rightmost two digits) is divisible by 4, so the number shall be divisible by 4.

(ii)   4, 6, 8, 9, 10 are prime numbers   ( )

Soln: 4, 6, 8, 10 are even numbers and they are divisible by 2, so they are not prime numbers, because prime numbers should be divisible only by themselves or by 1. False

(iii)   2 is a composite number   ( )

Soln: 2 has only two factors, 2 and 1, so it is a prime number. False

(iv)   15 is a prime number   ( )

Soln: 15 = 3 x 5, so it has factors 3 and 5 and hence 15 is not a prime number. False

(v) $(42)_7 = (30)_{10}$ ( )

Soln: $(42)_7 = 2 \times 1 + 4 \times 7^1 = 2 + 28 = (30)_{10}$, so correct.

(vi) $(23)_5 = (31)_4$ ( )

Soln: converting both sides of expression to the base of 10,

$(23)_5 = 3 \times 1 + 2 \times 5^1 = 3 + 10 = (13)_{10}$

$(31)_4 = 1 \times 1 + 3 \times 4^1 = 1 + 12 = (13)_{10}$

So, True.

(vii) $(24)_5 = (15)_{10}$ ( )

Soln: $(24)_5 = 4 \times 1 + 2 \times 5^1 = 4 + 10 = 14 = (14)_{10}$, so, false.

(viii) $(111)_2 = (7)_{15}$ ( )

Soln: Left hand side = $(111)_2 = 1 \times 1 + 1 \times 2 + 1 \times 2^2 = 1 + 2 + 4 = 7$

Right hand side = $(7)_{15} = 7 \times 1 = (7)_{10}$

LHS = RHS, so True.

Other Problems

13. Fill in the blanks after understanding the pattern:
(i)   2, 5, 10, 17, 26…….

Soln: second number – first number = 5-2 = 3

Third number – second number = 10- 5 = 5

Fourth number – third number = 17- 10= 7

Fifth number – fourth number = 26- 17= 9

So, we see that difference between consecutive terms is 3, 5, 7, 9,.., i.e increasing by 2 with every term.

So, sixth number – fifth number = 9 + 2 = 11

So, sixth number = 26 + 11 = 37

Ans.

(ii)   8, 12, 9, 16, 10, 20, 11………..

Soln: This is mixture of two series. One series through combination of numbers at odd places, 1st, 3rd, 5th, …and second series through combination of numbers at even places, i.e., 2nd, 4th, 6th …

Numbers at odd places: 8, 9, 10, 11. Each number is increasing by 1, so 9th number shall be 12.

Numbers at even places: 12, 16, 20, so numbers increasing by 4. So next number at 8th place is 20 + 4 = 24, Ans.

(iii)   8, 24, 27, 9, 27, 30, 10……..

Soln: series can be written like follows:

8, (8 x 3), (8 x 3 + 3), (8 + 1= 9), (9 x 3), (9 x 3 + 3), 10, so next term is (10 x 3 = 30). Ans. 30

(iv)   3, 6, 11, 18, 27, ……..

Soln: 2nd term – 1st term = 6-3 = 3

3rd term – 2nd term = 11- 6 = 5

4th term – 3rd term = 18 -11 = 7

5th term – 4th term = 27 – 18 = 9

So, difference between consecutive numbers is increasing by 2.

Accordingly, 6th term $- 5^{th}$ term $= 11$

Or, 6th term $= 5^{th}$ term $+ 11 = 27 + 11 = 38$, Ans.

    (v)   2, 6, 12, 20, 30, ...

Soln: difference between terms is 4, 6, 8 and 10. So difference between next term and 30 = 12. So next number = 30 + 12 = 42, Ans.

    (vi)   1, 4, 7, 10, 13,.......

Soln: Increasing by 3, so next number = 13 + 3 = 16.

    (vii)   53, 37, 27, 17,........

Soln:

    (viii)   2, 5, 8, 11, ...

Soln: each term is 3 more than previous term. So, next number = 11 + 3 = 14, Ans.

    (ix)   3, -7, -11, -15, -19,........

Soln: Next term decreasing from previous term by 4. So next number = -19- 4 = -23, Ans.

    (x)   2, 9, 28, 71,......

Soln: $2 = 1^3 + 1$

$9 = 2^3 + 1$

$28 = 3^3 + 1$

$65 = 4^3 + 1$

    (xi)   42, 37, 32, 27,......

Soln: 42, (42-5= 37), (37-5= 32), (32-5= 27), so next number = (27-5 = 22, Ans.

(xii)	7	14	12
	4	12	9
	6	24	....

Soln:

Series can be understood like follows:

7, 7 x 2 = 14, 7 x 2 – 2= 12

4, 4 x 3 = 12, 4 x 3 – 3= 9

6, 6 x 4 = 24, **6 x 4 – 4 = 20**

So, missing number = 20, Ans.

(xiii)	30	17	21
	18	15	27
	7	.....	30

Soln: series can be written like follows:

30	(30 + 21)/3	21
18	(18 + 27)/ 3	27
7	(7 + 30)/ 3	30

So, missing number is 37/3, Ans.

14. Fill in the blanks below:
(i)   -5 is ....than -1

Soln: -5 is having more magnitude but sign is negative, so it is less.

(ii)   (3+ 5) + 7 = 3 + (5 +7) explains the ....law of integers.

Soln: Associative

(iii)   If a and b are prime numbers then a + b is also a prime number. This law is known as ....law.- check
(iv)   Additive identity for positive integers is ....

Soln: Zero, because in whichever number it is added, the result of the addition remains same as the number.

15. Write answer of following in front of questions:
(i)   Sum of two integers is an integer is according to which property of integers....................

Soln:

(ii)   For rational numbers, multiplicative identity is...................

Soln: 1, because whichever number is multiplied by 1, the result of multiplication is that number itself.

(iii)   A (B + C) = AB + AC is as per....................law

Soln: Distributive law

(iv)   A x (B x C) = (A x B) x C, shows ....property of numbers.

Soln: Associative

16. 7 x (3 x 8) = (7 x 3) x 8, is based on ....law of product of numbers.

Soln: Associative.

17. 7 + (-13) = (-13) + 7, is as per ....law.

Soln: Commutative

18. Which is greater number, -4/5 or -7/9.ke

Soln: to compare two fractions, their denominators are made equal and then the fraction with greater numerator is greater.

For making denominators equal we calculate LCM of the denominator. LCM of 5 and 9 = 5 x 9 = 45, because they are prime to each other (no common factor).

Now making denominators of two fractions equal to the LCM.

-4/5 = (- 4 x 9)/ (5 x 9) = - 36/ 45

- 7/9 = (- 7 x 5) / (9 x 5) = -35/ 45

So, denominators are equal, now comparing numerators, - 35 is greater than -36, so fraction corresponding to -35, i.e. – 7/9 is greater.

Ans. -7/9

19. Express 1000 in powers of 10.

Soln: $1000 = 10 \times 10 \times 10 = 10^3$

20. What is sign of number got through raising negative numbers to odd number.

Soln: A negative number = (-1) x (positive number)

So, $(\text{negative number})^{\text{odd number}} = (-1)^{\text{odd number}} \times (\text{positive number})^{\text{odd number}}$

Now, any power of a positive number shall always be positive number and -1 multiplied odd number of times = -1 (check for 3, 5 and any other odd value).

So, $(-1)^{\text{odd number}} \times (\text{positive number})^{\text{odd number}} = - 1 \times (\text{positive number}) = \text{negative number}$.

Ans.

21. What is exponent (power), if 64 is written in powers of 2.

Soln: $64 = 2 \times 2 \times 2 \times 2 \times 2 \times 2 = 2^6$

So exponent is 6. Ans.

22. a(b+c) = ab + ac, shows.......................law

Soln: Distributive law

23. which one is greater: 3/5 or 4/5

Soln: 4 > 3

So, 4/ 5 > 3/5

Ans.: 4/5

24. which one is greater: 7/8 or 2/3

Soln: Making denominators equal.

LCM of 8 and 3 = 24

7/8 = (7 x 3) / (8 x 3) = 21 / 24

2/3 = (2 x 8) / (3 x 8) = 16/ 24

21 is greater than 16, so, 21/24 > 16/ 24, so, 7/8 > 2 / 3

Ans.: 7/8

25. $\{(5)^2\}^3$ is equal to:
      (i)   5 x 2 x 3   (ii) 25 x 3   (iii) $(5)^5$ (iv) $(5)^6$

Soln: $\{(5)^2\}^3 = (5)^2 \times (5)^2 \times (5)^2 = (5 \times 5) \times (5 \times 5) \times (5 \times 5) = 5 \times 5 \times 5 \times 5 \times 5 \times 5 = 5^6$, Ans. (iv)

26. ....multiplicative identity for natural numbers.

Soln: 1

27. Which numbers are represented by following:
      (i) $(2t2)_{12}$ (ii) $(5te)_{12}$ (iii) $(e5t)_{12}$

Soln:

      (i) $(2t2)_{12}$

Meaning of t is ten here.

$(2t2)_{12} = 2 \times 12^0 + 10 \times 12^1 + 2 \times 12^2$

$= 2 \times 1 + 10 \times 12 + 2 \times 144$

$= 2 + 120 + 288$

$= 410$, Ans.

      (ii) $(5te)_{12}$

Meaning of t is ten and e is eleven in numbers to the base 12.

So, $(5te)_{12} = 11 + 10 \times 12 + 5 \times 12 \times 12 = 11 + 120 + 720 = 851$, Ans.

28. In which system only (0, 1, 2) digits are used.

Soln: For a number system at base N, numbers up to N-1 are used. Accordingly the base of system using digits up to 2 shall be (2 + 1) = 3. Ans.

29. Write first four natural numbers.

Soln: 1, 2, 3, 4

30. Write first four even numbers.

Soln: 2, 4, 6, 8

31. Write first four odd numbers.

Soln: 1, 3, 5, 7

32. Write one prime number between 5 and 10.

Soln: 7

33. Express 0.12 as prime number.

Soln: $0.12 = 12/100 = 4 \times 3 / 25 \times 4 = 3 / 25$, Ans

34. What is additive inverse of 5.

Soln: Additive inverse means if additive inverse of the number is added to the number, the summation shall be zero.

Let it be x, so $5 + x = 0$, or $x = -5$, Ans

35. What is multiplicative inverse of -1/4.

Soln: -4, multiplication should be 1.

36. Which number is highest two digit prime number less than 42.

Soln: 41

37. Which number is with three

Soln:

38. $(4 + 13) = (13 + 4)$ is as per..........................law of addition of integers.

Soln: commutative law

39. Find the largest number of three digits which is divisible by 5.

Soln: The largest 3 digit number = 999

999 ÷ 5 has quotient of 199 and remainder of 4. If remainder of 4 is deducted then the number shall be the highest 3 digit number which shall be divisible by 5.

So, number = 999 – 4 = 995, Ans

40. Which even number is a prime number also.

Soln: 2

41. The smallest number of five digits is.........................

Soln: 10000

42. (LXVI) is equal to...........................

Soln: 50 + 10 + 5 + 1 = 66

43. Value of $(3)^3$ is equal to ....

Soln: $(3)^3$ = 3 x 3 x 3 = 27

44. Additive inverse of 8/3 is ....

Soln: Additive inverse will be just opposite in sign so that sum is zero. So additive inverse = -(8/3) = -8/3

45. Value of $\{(5)^2\}^3$ is.............................

Soln: 5^6

46. If 10000 is written as powers of 10, what is exponent (power).

Soln: 10000 = 10 x 10 x 10 x 10 = 10^4

So exponent is 4.

47. The smallest prime number between 1 and 10 is ....

Soln: 2

48. Write two odd prime numbers, whose sum is 32.

Soln: 32 = 3 + 29

Ans.: 3 and 29